P9-CFG-063

P242w

WOODPILE

WOODPILE

WRITTEN AND ILLUSTRATED BY PETER PARNALL

Macmillan Publishing Company   New York

Collier Macmillan Publishers   London

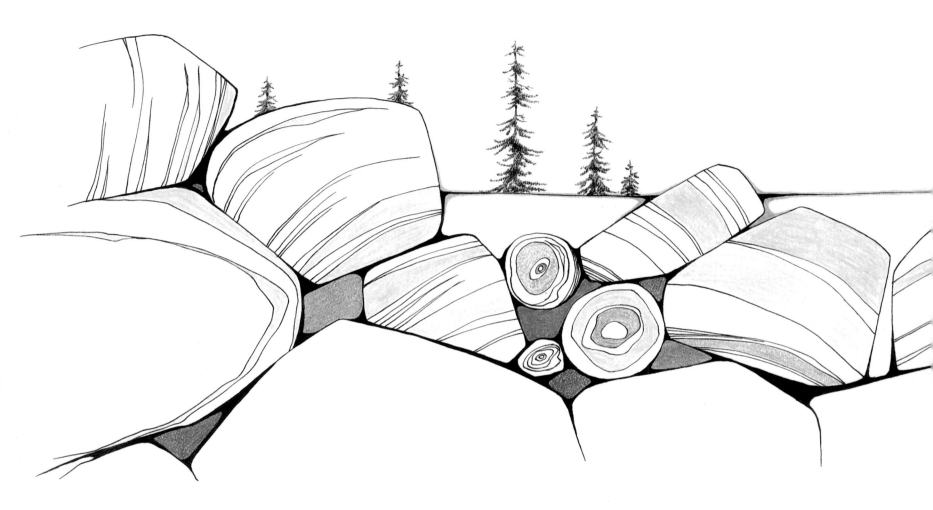

Macmillan Publishing Company, 866 Third Avenue, New York, NY 10022. Collier Macmillan Canada, Inc.
Printed and bound in Hong Kong   First American Edition   10 9 8 7 6 5 4 3 2 1
The text of this book is set in 14 point Baskerville. The illustrations are rendered in ink, pencil, and watercolor.
Library of Congress Cataloging-in-Publication Data
Parnall, Peter. Woodpile / written and illustrated by Peter Parnall. —1st ed.   p.   cm.
Summary: Depicts the various creatures that make a cozy home in the author's old pile of wood.
1. Woodpile fauna—Juvenile literature. [1. Woodpile animals.] I. Title.
QL49.P2575 1990 591.56′4—dc20 89-29322 CIP AC ISBN 0-02-770155-7

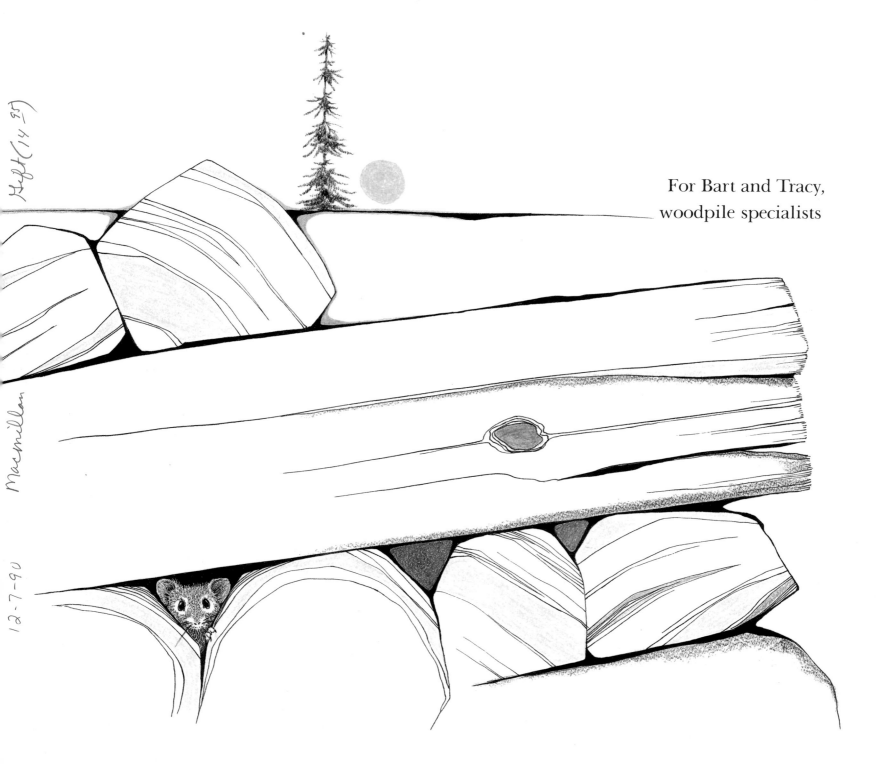

For Bart and Tracy,
woodpile specialists

My woodpile waits for frosty days. Days when leaves
turn red and gold, when green is just a memory and a wood
stove warms my bones. I see it there beyond my door—
maple, oak, birch, and ash, woven into a wooden world.
Apple, too. Small pieces cut from a limb that fell from a tree
planted by a soldier from the Civil War. He ate an apple
from that gnarly old tree. So did I, just last fall.

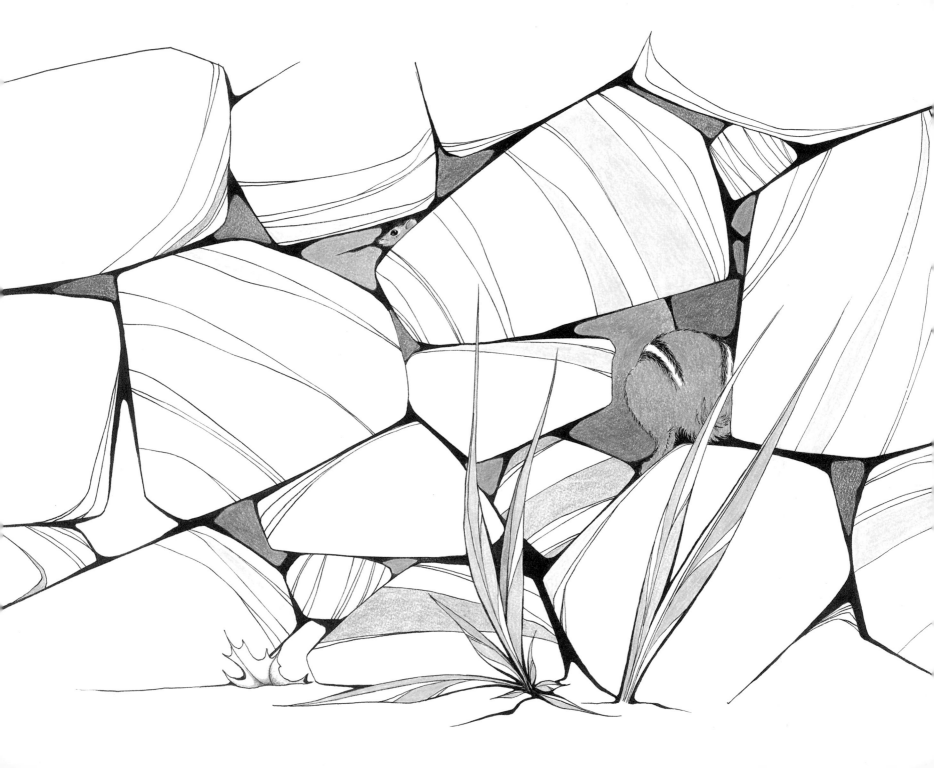

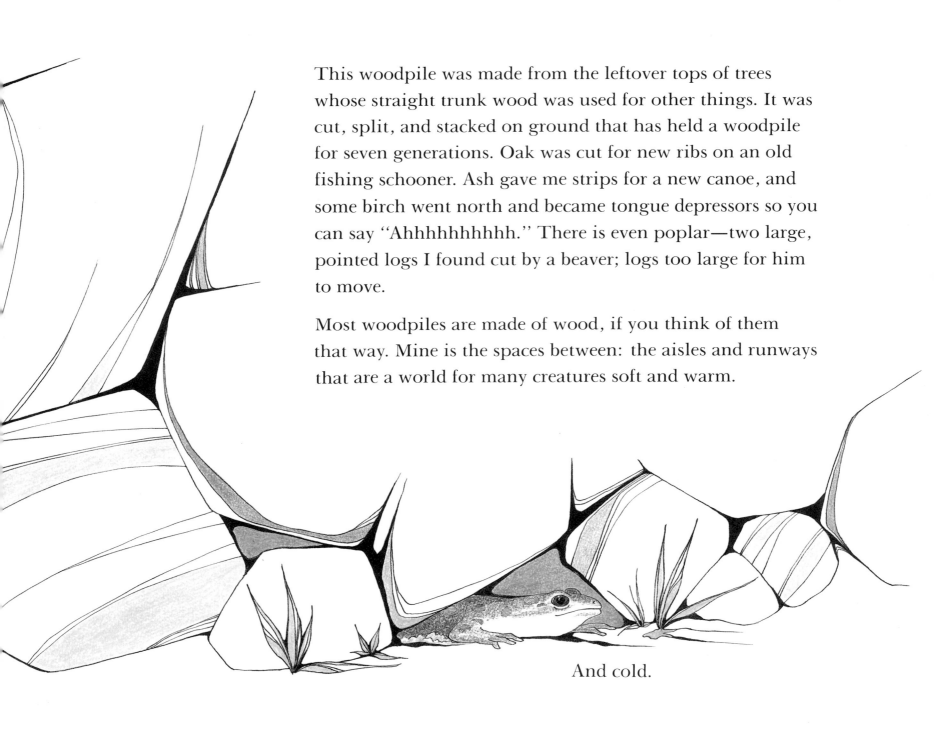

This woodpile was made from the leftover tops of trees whose straight trunk wood was used for other things. It was cut, split, and stacked on ground that has held a woodpile for seven generations. Oak was cut for new ribs on an old fishing schooner. Ash gave me strips for a new canoe, and some birch went north and became tongue depressors so you can say "Ahhhhhhhhhh." There is even poplar—two large, pointed logs I found cut by a beaver; logs too large for him to move.

Most woodpiles are made of wood, if you think of them that way. Mine is the spaces between: the aisles and runways that are a world for many creatures soft and warm.

And cold.

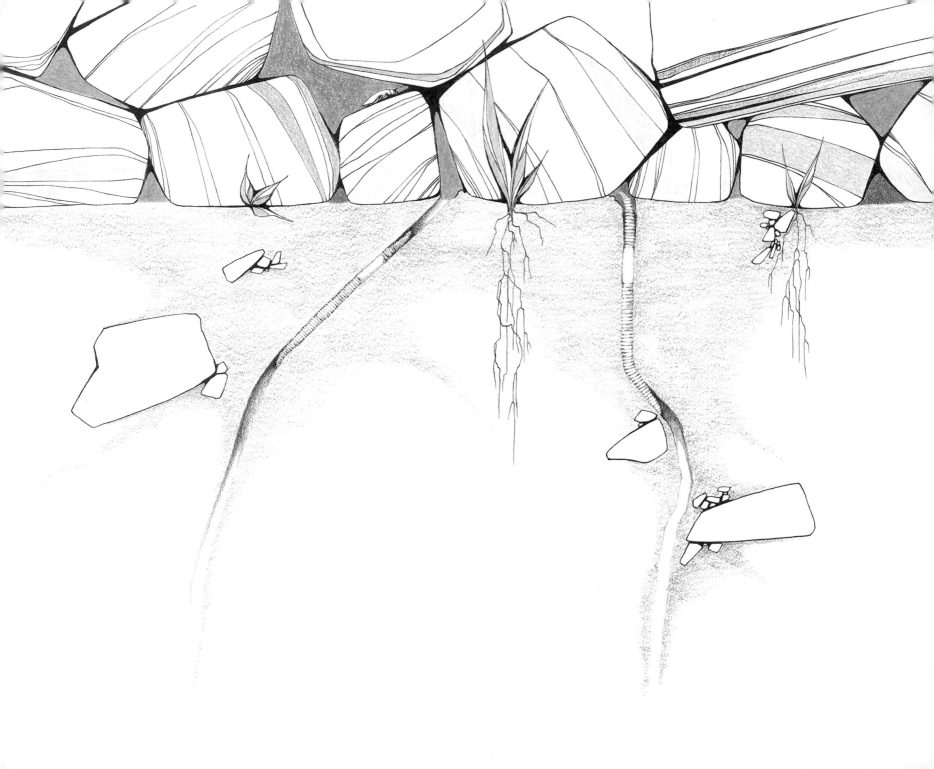

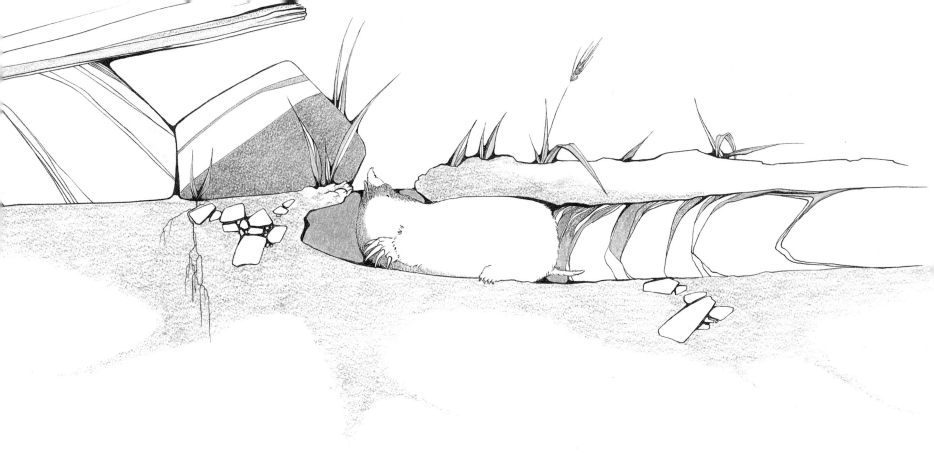

Beneath the pile the dirt is packed as hard as pavement, yet
Worm tunnels easily. Not Mole. When his path wandered to
the woodpile's base he burrowed to a solid dead end under
a cherry log cut from a tree that I made into a table for
flowers...daisies, mostly.

Over Mole's head, deep in a hollow beech log, Mouse arranges a nest of dried, soft treasures she has found on her nightly journeys. Her log came from a tree by the creek where wood ducks nest in beeches too old and hollow to withstand a winter's storm. When one does fall Skunk finds shelter within its bones.

Weasel darts nervously near Mouse's log, hunting. He cannot find her entrance: a tiny knothole out of sight.

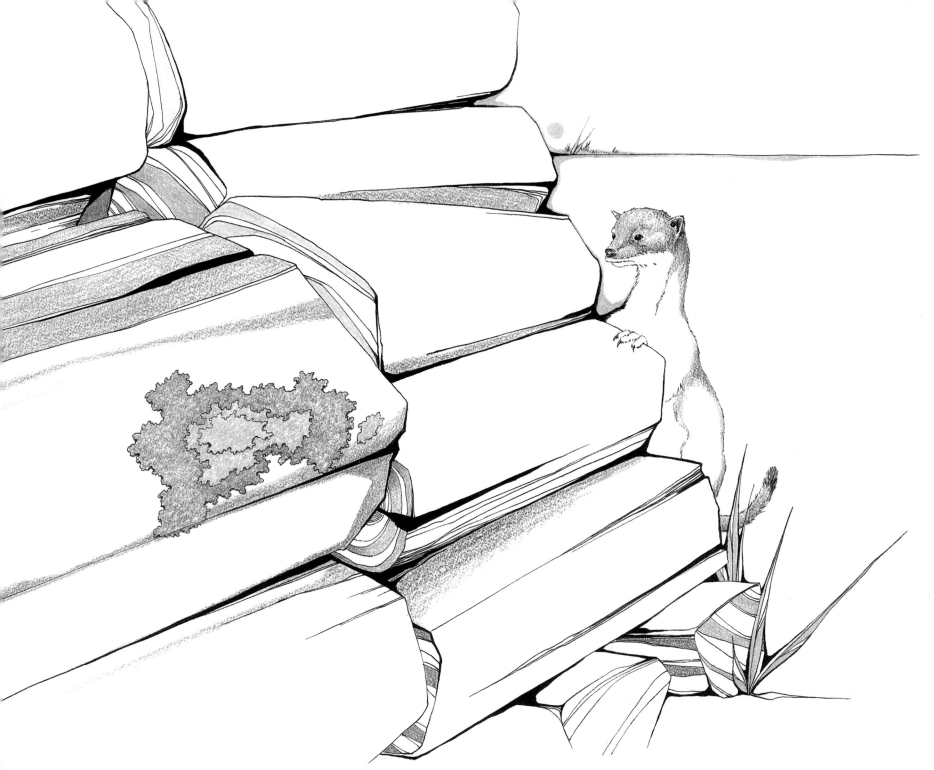

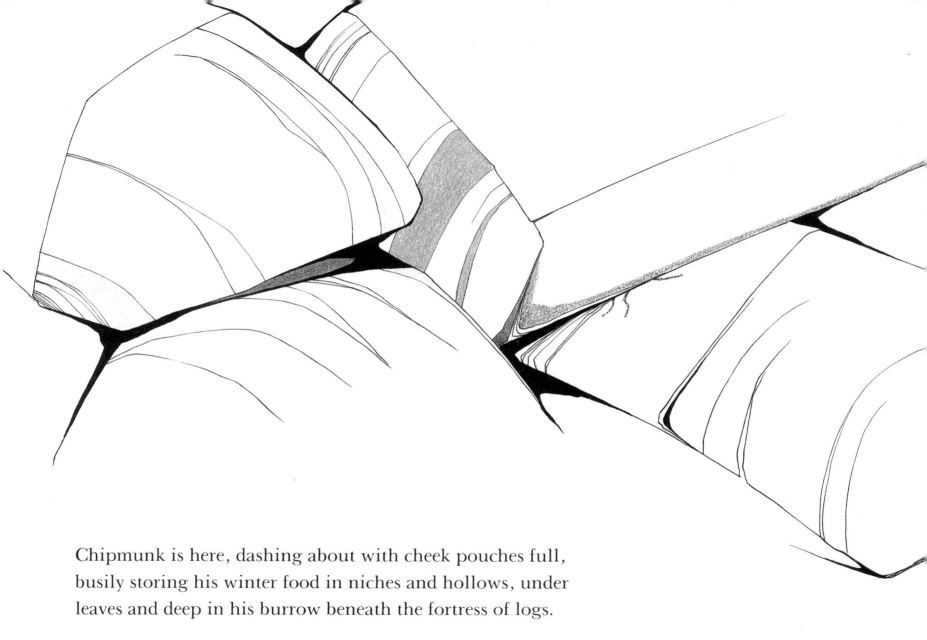

Chipmunk is here, dashing about with cheek pouches full,
busily storing his winter food in niches and hollows, under
leaves and deep in his burrow beneath the fortress of logs.

He knows when danger is lurking about and warns others
with a *chirrup* and a flick of his tail...

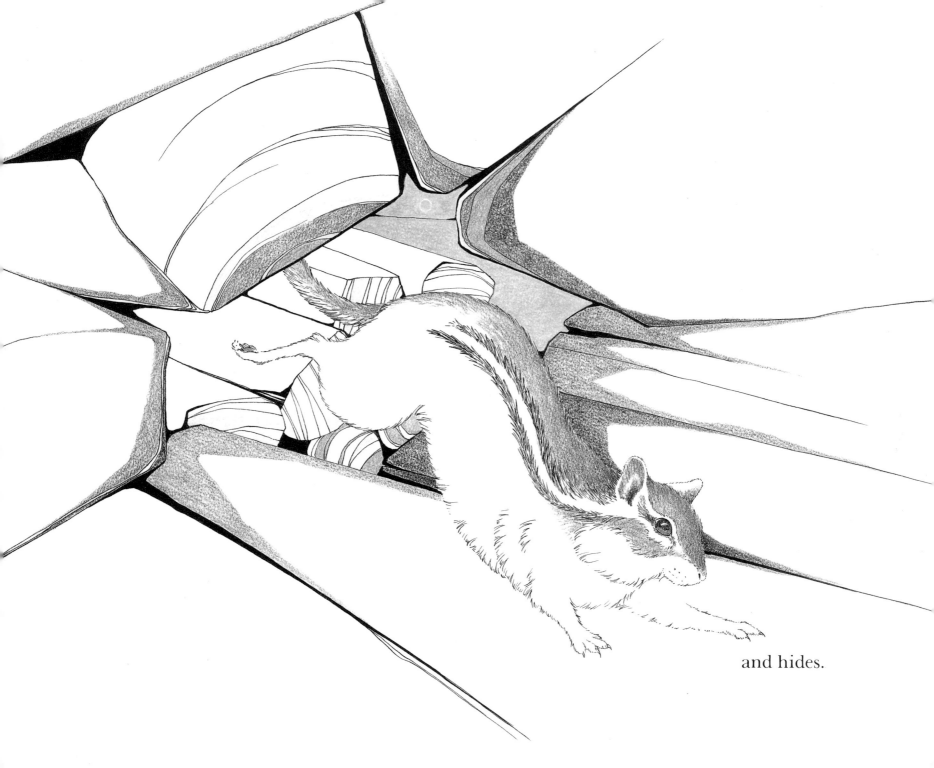

and hides.

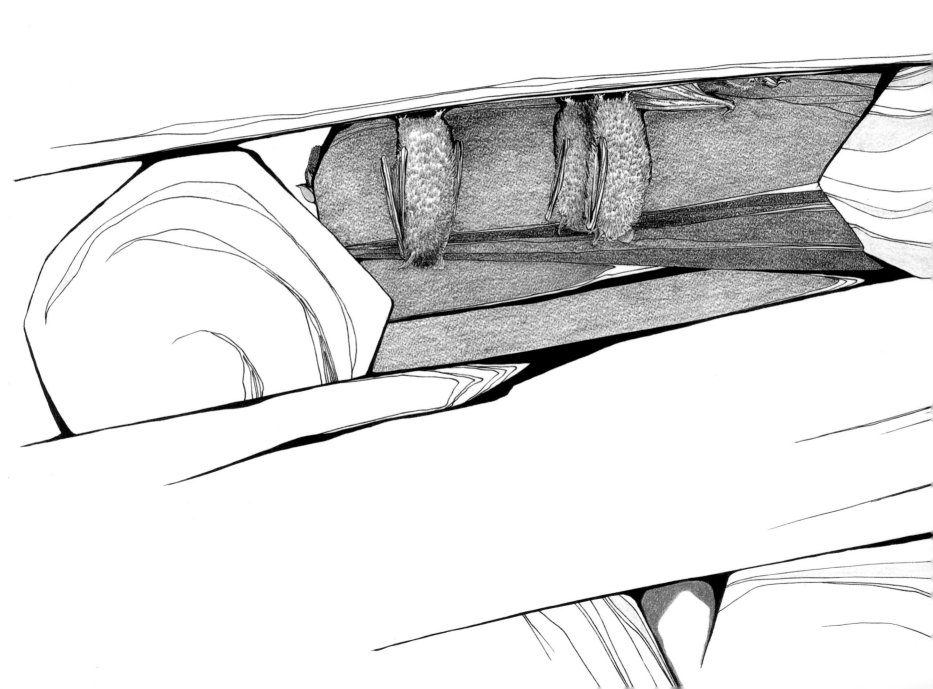

Some bats sleep here during the heat of the day, far from the barn where they were born. The barn roof cooks in the summer sun. It *bakes*.

If I were a bat I'd not live there at all, but here, clinging to the cool, dark bark of an old oak log surrounded by air that smells of sweet cherry and lilacs.

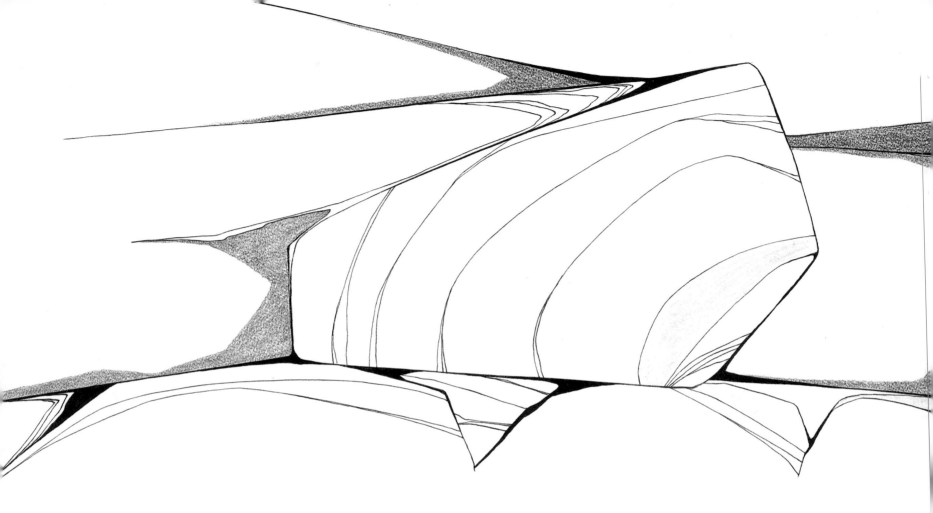

Underneath an overhanging maple log a wasp queen busies
herself strengthening her nest. She chews rotten wood from
maple and oak till it is fine enough to flow, and forms it into
thin mâché to build her paper home. When the sun climbs
high and creatures from Beyond seek shelter in the shadowy
pile, rising air keeps her home moist and cool.

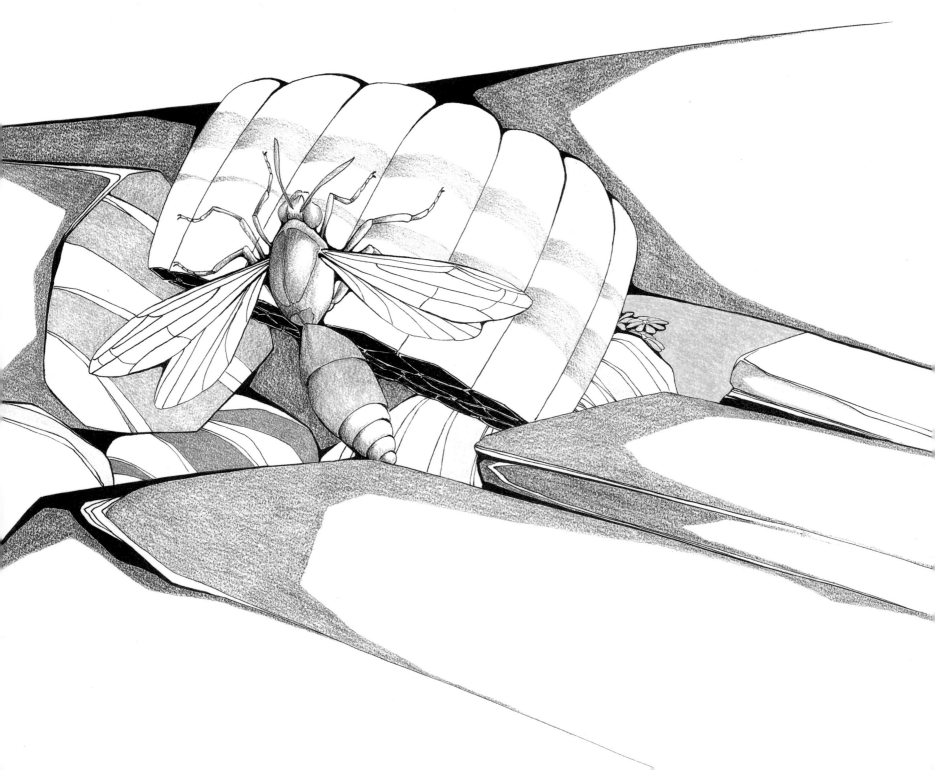

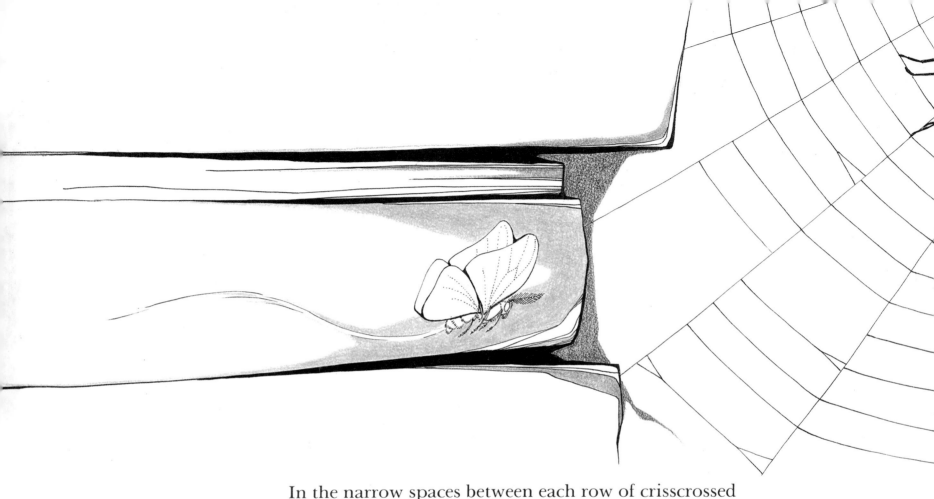

In the narrow spaces between each row of crisscrossed chunks spiders live and weave their webs to catch whoever dares fly into the shadowy alleys of wood. When Sun moves low, Mosquito, Moth, and flies of all kinds challenge the maze of sticky traps. Mosquito senses the warmth of Mouse, and in her haste falls prey to Spider's threaded puzzle.

Moth flutters by.

She bursts from the pile, her jerky flight becoming her armor as Owl drops silently from his lookout limb…and misses! For one hour a day during one week each year Screech Owl calls this his hunting ground. He sits just overhead on a maple limb, round gold eyes riveted to his wooden pantry. Many creatures could fall prey, like frogs and mice, but he likes moths.

Big, white moths.

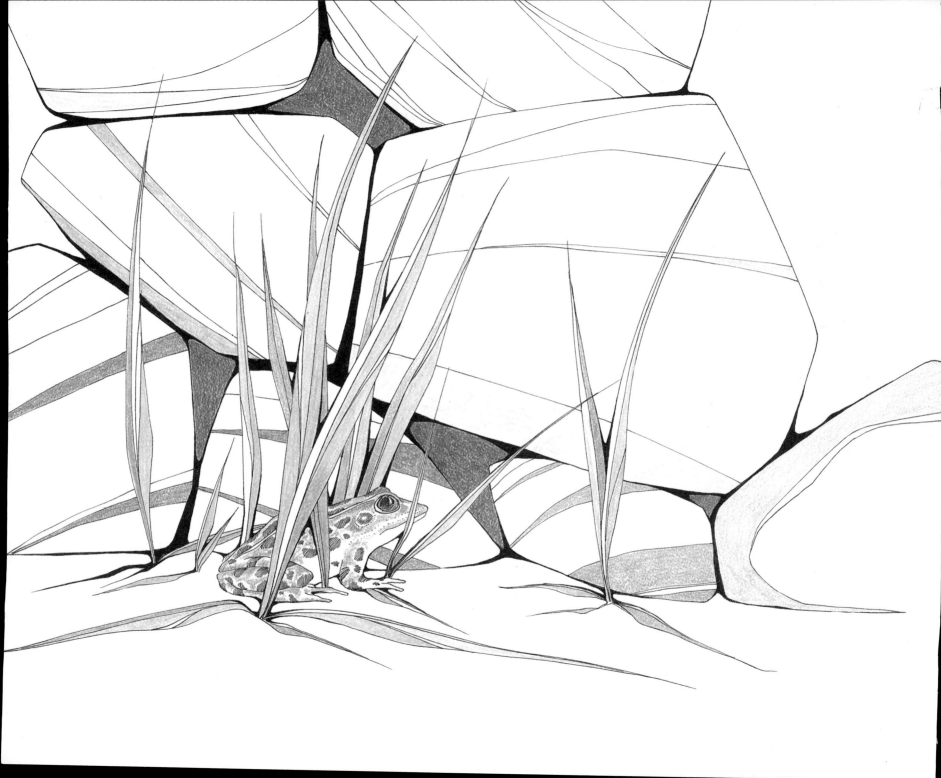

When Owl is near, Leopard Frog hides in tall grass that is too close to the pile to be mowed. Later he moves to the edge of the grass and waits for Mosquito, Fly, or anyone else small enough to pull in close with his sticky tongue. There too is Toad, half-buried in the refuse of the splitting axe, under bark chips, leaves, and wisps of dirt, waiting his chance for an evening's meal. I think Toad likes the smell of cherry wood in the rain. He is most often there, near the sweeter end of this wooden world.

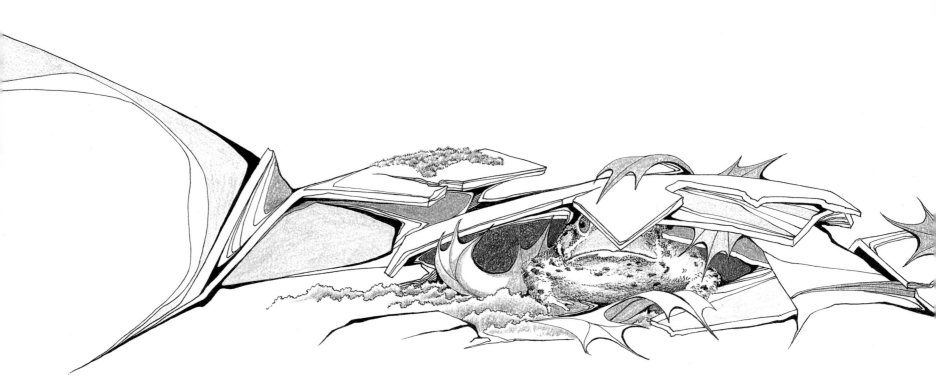

Skunk sometimes visits while on her nightly rounds. After digging for grubs in the lawn nearby she ambles over and turns over slabs of bark, looking for dessert...or Frog!

Phoebe owns a lookout log. She perches there on a piece of maple that is furry and green. It comes from a part of the marsh where lichens and moss grow so densely it seems a perfect world for gnomes and elves. The only gnome I've seen there is Porcupine. So far. It is Phoebe's log. No one else ever sits there.

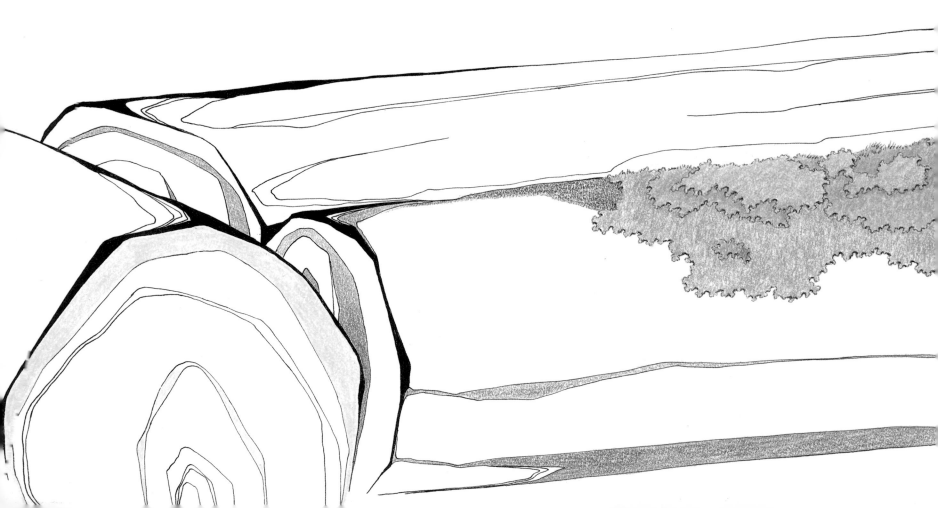

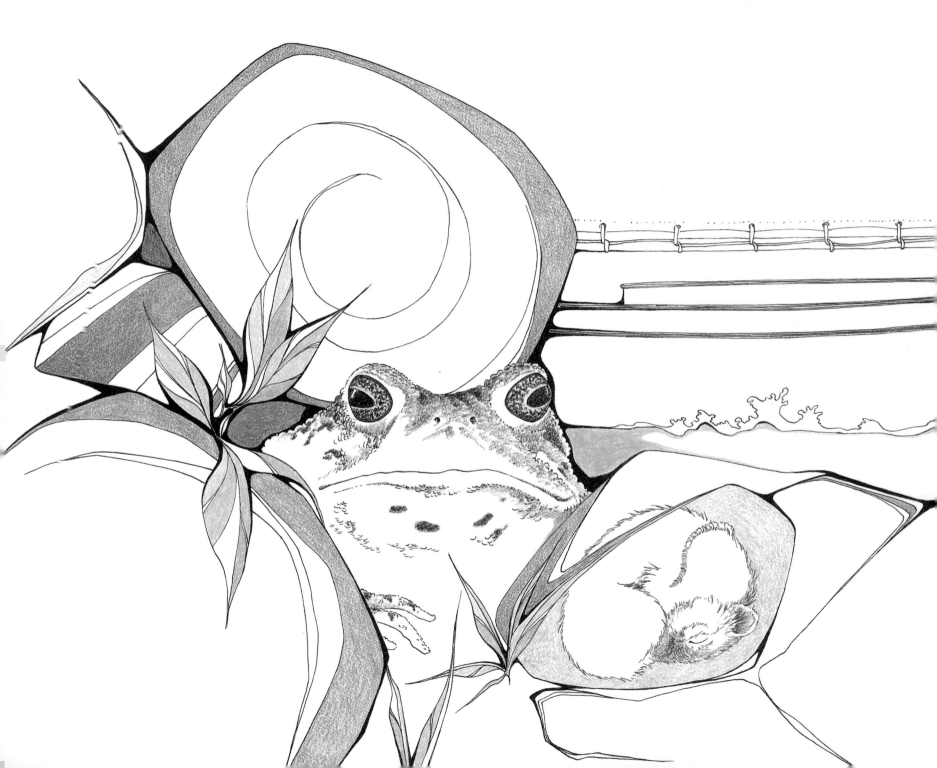

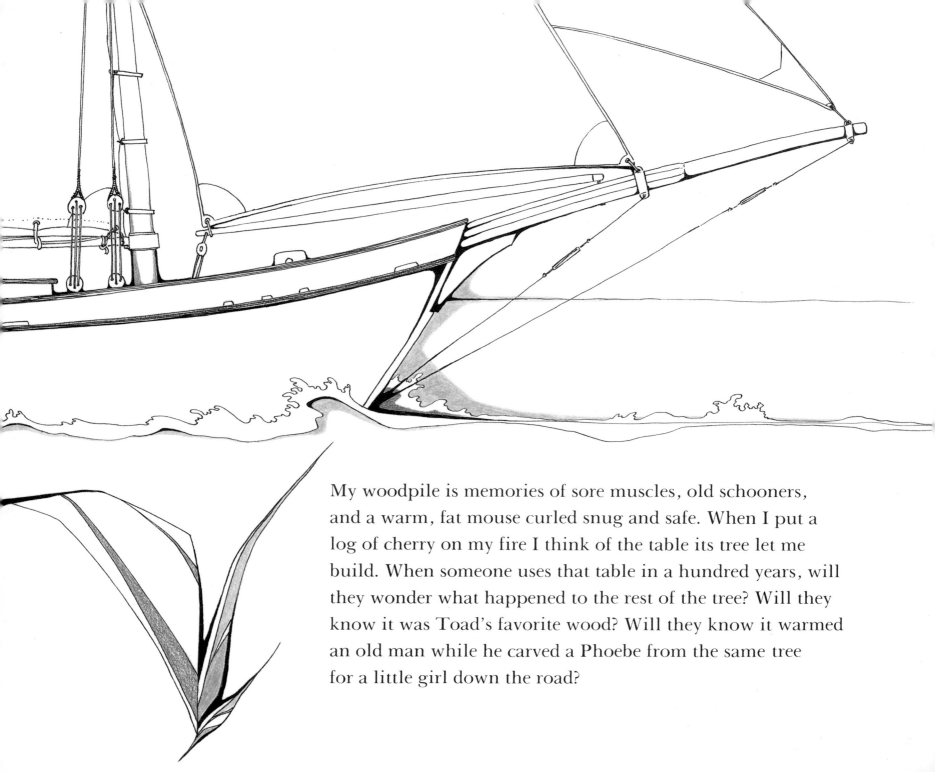

My woodpile is memories of sore muscles, old schooners, and a warm, fat mouse curled snug and safe. When I put a log of cherry on my fire I think of the table its tree let me build. When someone uses that table in a hundred years, will they wonder what happened to the rest of the tree? Will they know it was Toad's favorite wood? Will they know it warmed an old man while he carved a Phoebe from the same tree for a little girl down the road?

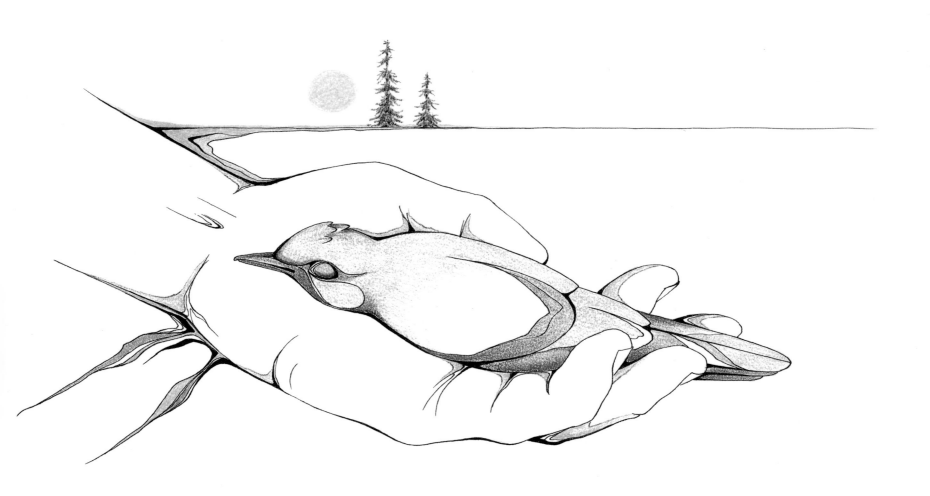

Maybe that little bird will help whoever holds it in their
hands be part of old schooners, cozy mice, owls, toads, and
an old pile of wood.

                                        And me.